Safety and Health Division

Core Programming Topic at the 2011 AIChE Annual Meeting

Minneapolis, Minnesota, USA
16-21 October 2011

ISBN: 978-1-61839-732-4

Printed from e-media with permission by:

Curran Associates, Inc.
57 Morehouse Lane
Red Hook, NY 12571

Some format issues inherent in the e-media version may also appear in this print version.

Printed by Curran Associates, Inc. (2012)

For permission requests, please contact AIChE
at the address below.

AIChE
3 Park Avenue
New York, NY 10016-5991

Phone: (203) 702-7660
Fax: (203) 775-5177

www.aiche.org

Additional copies of this publication are available from:

Curran Associates, Inc.
57 Morehouse Lane
Red Hook, NY 12571 USA
Phone: 845-758-0400
Fax: 845-758-2634
Email: curran@proceedings.com
Web: www.proceedings.com

TABLE OF CONTENTS

(232a) Process Hazards Emphasis In the New ABET Chemical Engineering Program Criteria

Jeffrey J. Siirola, AIChE, Kingsport, TN

This presentation will outline the US Chemical Safety Board concerns and recommendations that lead to the drafting of new ABET chemical engineering program criteria, the wording that is currently working its way through the ABET approval process, other ABET "editorial" changes being proposed for the chemical engineering program criteria, likely ultimate wording of the program criteria, and expected effective date of implementation. The intent of the Education and Accreditation Committee in drafting the new criteria wording will also be summarized.

See more of this Session: Integrating Safety Into the Chemical Engineering Curriculum - Where, What, and How (Part 1)
See more of this Group/Topical: Education

(232b) Promoting Acceptance of Process Safety Curriculum Requirements

Scott Berger, American Institute of Chemical Engineers, New York, NY

In 2009, the US Chemical Safety and Hazard Investigation Board (CSB) recommended to AIChE and ABET to develop requirements related to understanding safety hazards and how to prevent them. The resulting new ABET requirements are anticipated to take effect in 2012. It would not be unusual if some administrations of universities offering chemical engineering degrees were concerned or even upset about the new requirements.

In this short presentation, I will attempt to address the concern over the new requirements by discussing a broad range of benefits to the quality of the students' chemical engineering education and students' future employability, as well as some additional benefits that may accrue to the overall campus experience based on implementing the new requirements.

See more of this Session: Integrating Safety Into the Chemical Engineering Curriculum - Where, What, and How (Part 1)
See more of this Group/Topical: Education

(232c) Resources, Recommendations and Overcoming Challenges for Integrating Safety Into the Chemical Engineering Curriculum

Amy Theis, Fauske & Associates, LLC, Burr Ridge, IL

Interviews were performed with more than ten instructors and professors who are currently incorporating safety into the chemical engineering curriculum. These individuals described their methodology, resources and challenges in this effort. Excellent programs exist with courses dedicated to chemical process safety; some curricula also include a lab safety component. Implementation varies between schools; best practices and tips for effective teaching methods will be summarized. Examples of methods to overcome both explicit and perceived challenges will be discussed. Implementation strategies will be provided for universities with minimal safety experience as well as those that want to achieve a higher level of excellence. A relationship between industry and universities should exist to reinforce this commitment to safety; examples of this type of partnership will be presented.

See more of this Session: Integrating Safety Into the Chemical Engineering Curriculum - Where, What, and How (Part 1)
See more of this Group/Topical: Education

(232d) Educational Resources On Process Safety At the SaChE Website

Thomas O. Spicer, Ralph E. Martin Department of Chemical Engineering, University of Arkansas, Fayetteville, AR

Started in 1992, the Safety and Chemical Engineering Education Program (SaChE) is a cooperative effort between CCPS and engineering educators to provide teaching materials and programs for the enhancement of undergraduate and graduate students studying chemical and biochemical products and processes.

The SaChE Committee is comprised of representatives from academe, industry, and government in addition to AIChE staff. The SaChE Committee develops, reviews, and distributes classroom and training materials (SaChE Products) with the ultimate aim of safer operations. At its inception, funding for SaChE was provided from CCPS as well as membership fees from participating universities. As part of the Safety Scale-Up Partnership between CCPS and corporate sponsors, SaChE membership has been provided for all US and Canadian chemical engineering departments with AIChE student chapters.

Starting in 2003, SaChE Products were distributed by a web site hosted by the Ralph E. Martin Department of Chemical Engineering at the University of Arkansas (http://www.sache.org). Since the original development of the site, features have been added to further the goal of safety education including: Process Safety Beacon archive; Student Safety Certificate Program (largely based on the SaChE Products); Materials from Faculty Workshops sponsored by SaChE/CCPS; and links to related materials (such as the CSB Video Room).

The site also serves an administrative function by listing the SaChE Committee members in addition to a messaging service that facilitates email communication with Committee members and university and corporate contacts. Much of the material on the site can be viewed by all visitors. Some material on the site is restricted to registered users including the SaChE Products and printable editions of the Process Safety Beacon archive. This paper summarizes the resources available on the site and discusses statistics on worldwide usage gathered by Google Analytics.

See more of this Session: Integrating Safety Into the Chemical Engineering Curriculum - Where, What, and How (Part 1)
See more of this Group/Topical: Education

Continuing Our Journey to Bridge the Process Safety Gap Between Academia and Industry: Meeting the New ABET Process Safety Expectations

Bruce K. Vaughen, Cabot Corporation, Tuscola, IL

Thomas O. Spicer, Ralph E. Martin Department of Chemical Engineering, University of Arkansas, Fayetteville, AR

D. "Trey" Morrison, Exponent, Lisle, IL

James A. Klein, DuPont, Minneapolis, MN

David A. Rockstraw, Department of Chemical Engineering, New Mexico State University, Las Cruces, NM

ABSTRACT

This paper discusses a proposed organizational framework for SAChE products and reference materials that will help meet the new Chemical Process Safety (CPS)-related instructional requirements. The goals of this paper include:

1) To show how the current CPS products, certificates and references correspond to industry's Process Safety Management (PSM) efforts,
2) To provide a summary of the latest efforts for organizing and displaying the products, safety certificates and CPS-related resources into a curriculum-based framework,
3) To describe a matrix that links available SAChE materials to a typical chemical engineering curriculum, and
4) To show a couple of examples of how these CPS products (including the SAChE Process Safety Certification Program) can be incorporated into a typical chemical engineering curriculum to meet the proposed changes from ABET.

The paper concludes with a brief description of future SAChE-related efforts to help us reduce our process safety-related risks.

BACKGROUND

The Safety and Chemical Engineering Education (SAChE) committee was created in 1992, in part, to help provide instructors with practical Chemical Process Safety (CPS)-related teaching materials (aka "products") from industry [1]. One of SAChE's goals is to help instructors who don't have extensive process safety experience prepare and teach CPS lessons in their classes. A stronger emphasis on CPS in the chemical engineering curriculum will help reduce industry's process safety risk helping prevent major incidents such as the explosion and fatalities that occurred at T2 Laboratories in 2007 [2]. The process safety-related gap between education and application can be visualized in Figure 1 with a broken "bridge over troubled waters" of potential fires, explosions, runaway reactions and toxic releases.

Currently, the SAChE Committee has overseen the creation of more than 90 different products, the development of the safety certificate program (six certificates are currently available for students to complete), and provided a clearinghouse for many useful links to other resources [1]. Now that SAChE has established its strong resource base of teaching materials, we believe it is time to enhance the SAChE product's organizational framework to help instructors more easily locate materials that meet the AIChE/CCPS Process Safety Expectation Guidelines. The objective of making SAChE materials more accessible to faculty will facilitate the response of academia to the proposed 2011-12 ABET Chemical Engineering Program-Specific Criteria to "address the hazards associated with these processes," as is recommended by AIChE and the CSB [2, 3, 4].

A PROPOSED SACHE FRAMEWORK – THE SACHE LIBRARY

The proposed SAChE Library – an organizational framework - is designed to help chemical engineering instructors find course-specific materials for their use that incorporate the elements of process safety. This section of the paper describes the development of the "SAChE Library" in four parts:

1) SAChE's Link to Process Safety Management (PSM)
2) SAChE's CPS Materials - A New Organizational Scheme
3) A Matrix Linking the SAChE CPS Materials to the ChE Curriculum
4) Examples for Integrating SAChE CPS Materials in a ChE Curriculum.

As is noted above, the SAChE CPS-related materials have been created to help instructors prepare lessons for their particular chemical engineering course (e.g., thermodynamics, kinetics, design, etc.). The SAChE Library – its organizational scheme - is designed to help instructors find their course-specific materials quickly.

Part 1) SAChE's Link to Process Safety Management (PSM)

A useful image that helps describe a "PSM System" is shown as a PSM Wheel in Figure 2 [5]. The PSM System's three parts are as follows:

I.	PSM Programs (its 12 PSM Elements)	The Spokes
II.	Information Technology Systems (ITS)	The Hub
III.	Operational Discipline (OD)	The Tire

This system is designed to reduce the risks associated with fires, explosions, runaway reactions and toxic releases when handling or processing hazardous materials. By preventing the Loss of Containment (LOC) of flammable or toxic materials, PSM Programs help protect the safety and health of people, the environment, the facilities and the business.

As is shown in Figure 2, Information Technology Systems (ITS) represent the hub of the PSM Wheel. Information Systems include both paper and electronic systems that help communicate PSM-related information between the spokes. The communication may be between equipment, between people or be interfaces between equipment and people. These systems may be simple or complex, local or corporate. They may help document corporate or legal requirements and may require different levels of authorization. PSM risks are minimized when effective communication exists between PSM Elements, when learnings are shared, and when people document, store, maintain and use the process information.

The benefits for effectively applying PSM Programs focus on this concept: with fewer leaks, we reduce the potential for injuries, process safety events, environmental harm, equipment failure, poor yields and poor quality. Hence, PSM Programs help reduce potential disability costs, help reduce potential air, land and water regulatory permit violations and fines, help reduce unexpected production shutdowns and capital equipment replacement costs. A list of the twelve PSM Elements is shown in Figure 3. The details for each PSM Element are beyond the scope of this paper and are described elsewhere [5]. The goal of a robust PSM System is to keep us and our environment healthy so that our neighbors, our customers and our stockholders are happy.

Part 2) SAChE's CPS Materials - A New Organizational Scheme

Currently, the SAChE Committee has overseen the creation of more than 90 different products [1]. Now that there are many educational materials and references, discussions have focused on how best to "frame" these efforts for an efficient "search" for instructors. Prior work has noted reorganizing the existing products within "capsules" designed for specific chemical engineering courses [6]. Hopefully, the efforts noted in this paper capture subsequent discussions about SAChE's Modules and its website.

Note that the synergistic effects of many educators and representatives from industry work together, helping the student better understand different aspects of chemical engineering during their educational experience. The SAChE Committee conducts workshops, in conjunction with representatives from industry, where faculty can visit an industrial facility and get hands-on experience with the application of process safety principles and practices. Note that the approach to incorporate industrial-specific educational materials and examples into existing courses is not unique to the process safety field, as other "modules" designed by experts for use by instructors exist in other fields like sustainability and green engineering [7], financial risk analysis [8] and the use of computers [9], as well.

<u>Part 3) A Matrix Linking the SAChE CPS Materials to the ChE Curriculum</u>

A Matrix linking courses in the Chemical Engineering Curriculum to the PSM System is shown in Figure 4. The rows of this matrix list typical courses in a chemical engineering curriculum, including both fundamental science and mathematics courses as well as chemical engineering specific courses. The columns of this matrix capture the parts of the PSM System shown in Figure 2 and listed in Figure 3. The SAChE Link correlates to the "Design" aspects of a robust PSM System: Process Safety Information (PSI), Process Hazards Analyses (PHA) and Management of Change (MOC). The specific link at the "Design" stage for process safety between education (SAChE) and application (PSM) is noted in Figure 3, shaded with respect to the other PSM Elements in Figure 4 and shown schematically as a "bridge" in Figure 5.

<u>Part 4) Examples for Integrating SAChE CPS Materials in a ChE Curriculum</u>

There are many ways to incorporate the elements of process safety into existing chemical engineering courses without having to create a dedicated process safety-related course within an already overcrowded curriculum. The objective of the new program-specific ABET criteria for chemical engineering can be met, in part, by integrating SAChE CPS Modules into existing courses. The following examples show how these CPS educational materials can be incorporated in the senior design course and how these materials are integrated across an engineering curriculum at New Mexico State University.

Example 1: One approach to meet the ABET expectations has been described in detail in prior work [10]. A SAChE Module, designed to help meet the new ABET expectations, helps bridge the gap between academia and industry by incorporating existing SAChE Modules into the senior design course: "Safety Guidance for Design Projects [11]". The Senior Design Project Block Diagram - a design course "Road Map" - is shown schematically in Figure 6. This Road Map can be used to help the students meet and document the new ABET process safety-related expectations.

Example 2: At New Mexico State University, students must complete seven SAChE Modules to meet their graduation requirements – it is required as a part of their degree. The SAChE Modules are assigned as stand-alone exercises, with "certification" in the assigned modules accounting for approximately 5% of course grade. The course examinations/quizzes include questions drawn from SAChE Module content. In addition, the AIChE Student Chapter offers viewings of the CPS-related videos followed by Q&A sessions with pizza.

The learning modules serve as independent, stand-alone educational assignments for students to complete outside of the classroom and independent of the lecture. These modules contain a bulleted list of learning objectives that permit an instructor to quickly identify relevant content suitable to in-class assessment of student learnings. The Modules are targeted for different education levels, from freshmen through seniors, to allow for easy integration of content across the entire student experience. Hence, over time, the students are immersed in the chemical industry's corporate safety philosophy.

OUR PATH FORWARD

Where we are today: Process Safety Management (PSM) is defined as the principles and systems applied to identify, understand and control process hazards to reduce process safety risks and protect the safety and health of people, neighbors and the environment [see reference 12 for a recent CCPS overview of PSM]. As risk managers, we focus on the hazards of our processes that may lead to a major Loss Of Containment (LOC), resulting in fires, explosions, runaway reactions and/or toxic releases. Our risk reduction equation, shown in Figure 7 and described in detail elsewhere [13, 14], is used as the basis for reducing our process safety risks. And as engineers across the globe, we "hold paramount the safety health and welfare of the public and protect the environment [15]," we "must be dedicated to the protection of the public health, safety, and welfare [16]," and we "work to enhance the welfare, health and safety of all whilst paying due regard to the environment and the sustainability of resources [17]."

For decades, experts across the globe have pooled and shared their experiences to help apply the principles of process safety to reduce our process safety risks [1, 9, 18, and 19]. There are visible risk reduction efforts, such as meeting the criteria of Responsible Care® [20], and invisible efforts simply performed and undocumented since they are "the right thing to do."

Where we may be going: The authors hope that these visible SAChE-related efforts contribute to bridging the process safety-related gap between academia (education) and industry (application), as is shown in Figure 8, helping reduce our process safety-related risks. We hope that the SAChE Modules are used to help ensure adequate education – combining both training and skills-based testing - meeting our specific regulatory process safety-related requirements to transform raw materials safely and efficiently into products, as well [e.g., 21, 22, 23, 24, and 25]. Future SAChE Committee efforts will include identifying which new teaching modules need to be created or which of the existing modules will need to be reviewed and updated. Future efforts will include the SAChE Website and system "scalability" – i.e., its sustainability – to ensure that accurate and up-to-date information is readily available to instructors.

REFERENCES

1. The Safety And Chemical Engineering Education (SAChE) Website, An AIChE Alliance, http://www.sache.org/ (accessed 21-August-2011).

2. The US Chemical Safety Board (CSB) T2 Laboratories Report, http://www.csb.gov/ (accessed 21-August-2011): completed investigations, T2 Laboratories Inc. Reactive Chemical Explosion, December 19, 2007.

3. Gunaratnam, Rachael T., "CSB Recommendations From the T2 Laboratories Inc. Investigation," paper presented in Session 339, "Integrating New Safety Requirements Into the Chemical Engineering Curriculum" Session, abstract 339a published in the Proceedings from the 2010 AIChE Annual Meeting, Salt Lake City, UT, November, 2010.

4. Willey, Ronald J., H. Scott Fogler and Michael B. Cutlip, "The Integration of Process Safety Into a Chemical Reaction Engineering Course: The Review of the T-2 Incident," paper presented in Session 339, "Integrating New Safety Requirements Into the Chemical Engineering Curriculum" Session, abstract 339b published in the Proceedings from the 2010 AIChE Annual Meeting, Salt Lake City, UT, November, 2010.

5. Vaughen, Bruce K., "A Process Safety Management (PSM) Overview," Safety and Chemical Engineering Education (SaChE), SaChE Products (accessed 21-August-2011), http://sache.org/products.asp, 2012.

6. Morrison, Delmar (Trey) R. and Russell A. Ogle, "Developing Process Safety Capsules for the Chemical Engineering Classroom," paper presented in Session 339 "Integrating New Safety Requirements Into the Chemical Engineering Curriculum" Session, published in the Proceedings from the 2010 AIChE Annual Meeting, Salt Lake City, UT, November, 2010.

7. Seay, Jeffery, Richard Rezek and Luke Richardson, "Developing Sustainable Engineering Modules for the Chemical Engineering Curriculum," paper presented in Session 405 "Incorporating Green Engineering and Sustainability Into the Curriculum," abstract 405e published in the Proceedings from the 2010 AIChE Annual Meeting, Salt Lake City, UT, November, 2010.

8. Papavassiliou, Dimitrios V., "Introducing Risk Analysis in a Design I Course," paper presented in Session 247 "Steal This Course!," abstract 247a published in the Proceedings from the 2010 AIChE Annual Meeting, Salt Lake City, UT, November, 2010.

9. Computer Aids for Chemical Engineering Education (CACHE), http://cache.org/ (accessed 21-August-2011).

10. Vaughen, Bruce K., "A SAChE Module Designed to Bridge Process Safety's Troubled Waters: Meeting the New Academic Process Safety Requirements," paper presented in Session 339 "Integrating New Safety Requirements Into the Chemical Engineering Curriculum" Session, published in the Proceedings from the 2010 AIChE Annual Meeting, Salt Lake City, UT, November, 2010. Submitted for review to *Chemical Engineering Education*, July 2011.

11. Vaughen, Bruce K., "Safety Guidance for Design Projects," Safety and Chemical Engineering Education (SaChE), SaChE Products, http://sache.org/products.asp, 2011 (accessed 21-August-2011).

12. Sepeda, A. L., "Understanding Process Safety Management," Chemical Engineering Progress, August 2010, Vol 106, No. 6, pp. 26-33.

13. Klein, James A., and Bruce K. Vaughen, "Implementing an Operational Discipline Program to Improve Plant Process Safety," Chemical Engineering Progress (CEP), The American Institute of Chemical Engineers (AIChE), June 2011, pp. 48-52.

14. Vaughen, Bruce K., and James A. Klein, "Improving Operational Discipline to Prevent Loss of Containment Incidents," Process Safety Progress (PSP), The American Institute of Chemical Engineers (AIChE), in press.

15. The American Institute of Chemical Engineers (AIChE), Code of Ethics, http://www.aiche.org/About/Code.aspx (accessed 21-August-2011).

16. The National Society of Professional Engineers (NSPE), Code of Ethics, http://www.nspe.org/Ethics/CodeofEthics/index.html (accessed 21-August-2011).

17. The Royal Academy of Engineering, in collaboration with Engineering Council (UK) and other societies, including the Institution of Chemical Engineers (IChemE): "Statement of Ethical Principles," (accessed 21-August-2011), http://www.raeng.org.uk/societygov/engineeringethics/principles.htm

18. The Center for Chemical Process Safety (CCPS), American Institute of Chemical Engineers (AIChE), http://www.aiche.org/CCPS/index.aspx (accessed 21-August-2011).

19. European Process Safety Centre (EPSC), http://www.epsc.org/ (accessed 21-August-2011).

20. American Chemistry Council (ACC), Responsible Care ® Management Systems and Certification, http://responsiblecare.americanchemistry.com/Responsible-Care-Program-Elements/Management-System-and-Certification/default.aspx (Accessed 21-August-2011).

21. Occupational Safety & Health Administration (OSHA), Standard, OSHA 29 CFR 1910.119, Process safety management of highly hazardous chemicals, http://www.osha.gov/pls/oshaweb/owadisp.show_document?p_table=STANDARDS&p_id=9760 (Accessed 21-August-2011).

22. Environmental Protection Agency (EPA), The Risk Management Plan (RMP) Rule, http://www.epa.gov/osweroe1/content/rmp/index.htm (Accessed 21-August-2011).

23. Health and Safety Executive (HSE), Control of major accident hazards (COMAH), http://www.hse.gov.uk/comah/index.htm , (Accessed 21-August-2011).

24. European Commission (EC), Environment, Chemical Accidents (Seveso II) - Prevention, Preparedness and Response http://ec.europa.eu/environment/seveso/index.htm, (Accessed 21-August-2011).

25. Guidelines for Process Safety Management, AQ/T3034-2010, Effective: 2011-May -01. The Guideline is the first standard regarding the Process Safety Management (PSM) in China (accessed 21-August-2011) "translate this page!" http://www.chemicalsafety.org.cn/News_Minute.aspx?newsid=8068

Figure 1

The Bridge Over Process Safety's Troubled Waters:

The Gap Between Education and Application

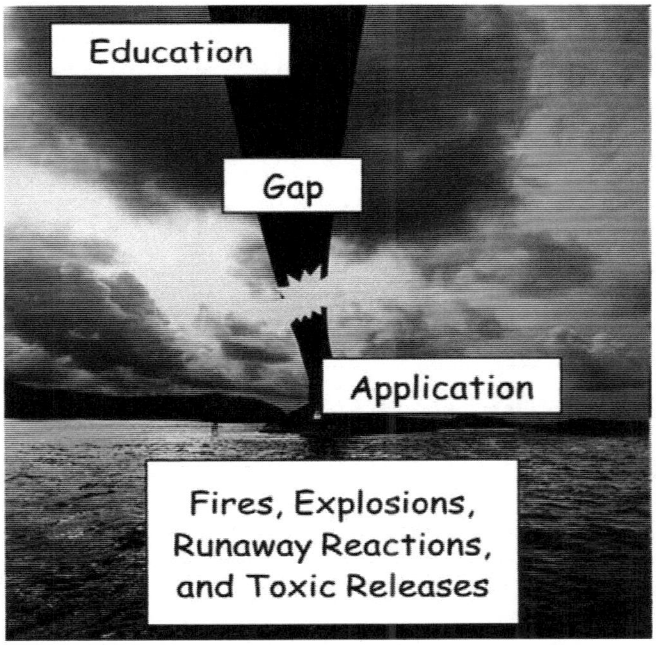

Figure 2

The Process Safety Management (PSM) System:

Image of a Wheel

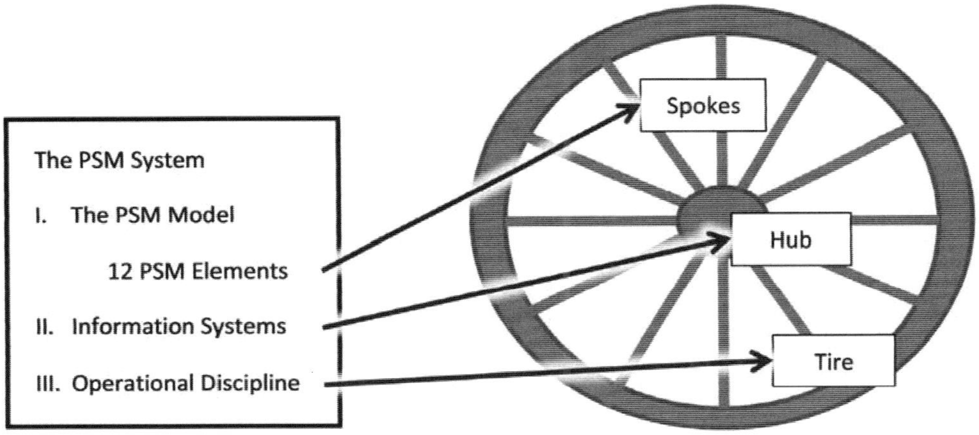

Figure 3

The Process Safety Management (PSM) System

I	The PSM Model	PSM
01	Process Safety Information	PSI
02	Process Hazards Analysis	PHA
03	Management of Change	MOC
04	Pre-Startup Safety Reviews	PSSR
05	Plant Integrity and Reliability	PIR
06	Written Procedures	WP
07	Safe Work Practices	SWP
08	Contractors and Visitors	C&V
09	Emergency Management & Planning	EMP
10	Training and Performance	T&P
11	Auditing	AUD
12	Investigations	INV
II	Information Systems	IS
III	Operational Discipline	OD

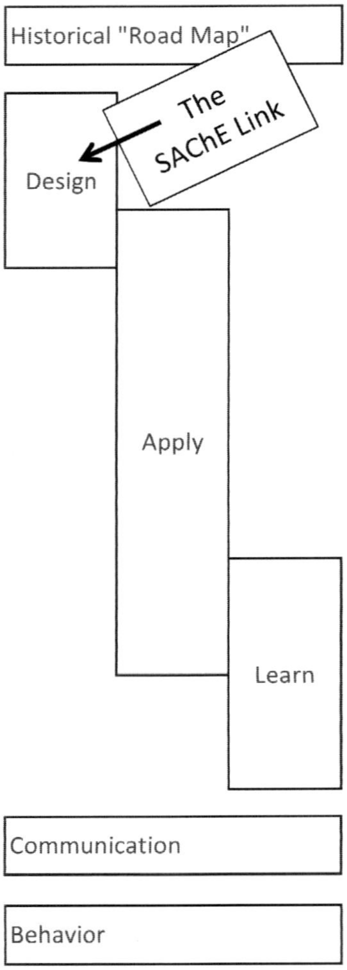

Figure 4

A Matrix Linking The Chemical Engineering Curriculum to the PSM System

		The PSM System													
		I. PSM Model												II.	III.
		Design	Apply									Learn			
		01	02	03	04	05	06	07	08	09	10	11	12	ITS	OD
Chemical Engineering Courses		PSI	PHA	MOC	PSSR	PIR	WP	SWP	C&V	EMP	T&P	AUD	INV		
Science and Mathematics Foundation	Mathematics	√	√	√		√						√	√	√	
	Engineering: Basic, Computers	√												√	
	Chemistry: Basic	√													
	Chemistry: Structure and Reactivity and Lab	√	√	√	√	√	√	√	√	√	√	√	√		√
	Chemistry: Intro to Quantum	√													
	Physics with Lab	√	√	√	√	√	√	√	√	√	√	√	√		√
	Biology/Life Science	√													
	Material Science	√													
	Other	√													
Chemical Engineering Specific	Material and Energy Balances	√				√									
	Thermodynamics	√													
	Fluid Mechanics	√													
	Mass and Heat Transfer	√													
	Separation Processes	√												√	
	Reaction Engineering and Design	√		√											
	Chemical Engineering Laboratory	√	√	√	√	√	√	√	√	√	√	√	√	√	√
	Process Dynamics and Control	√				√								√	
	Chemical Engineering Process Economics	√	√	√		√		√	√	√	√				
	Chemical Process Simulation and Design	√	√	√								√		√	

Note: Process Safety Information is required for all PSM Elements!

Figure 5

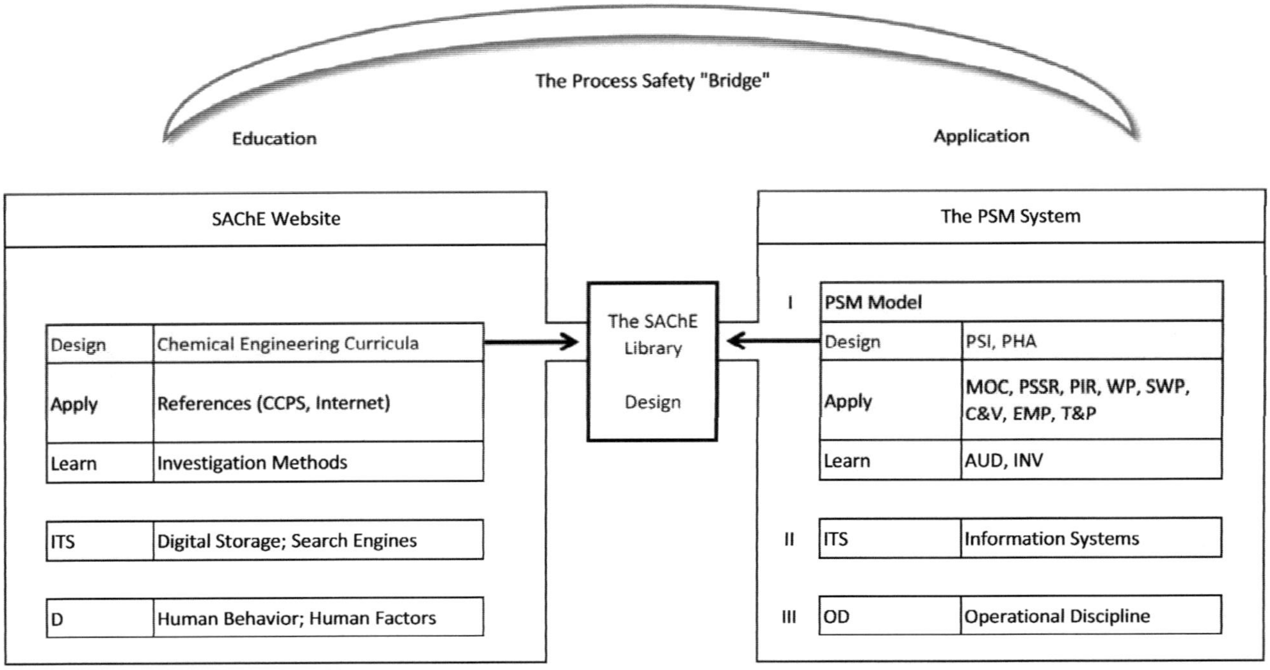

Figure 6

The Design Team Project Road Map

The steps addressing and documenting process safety within a design project

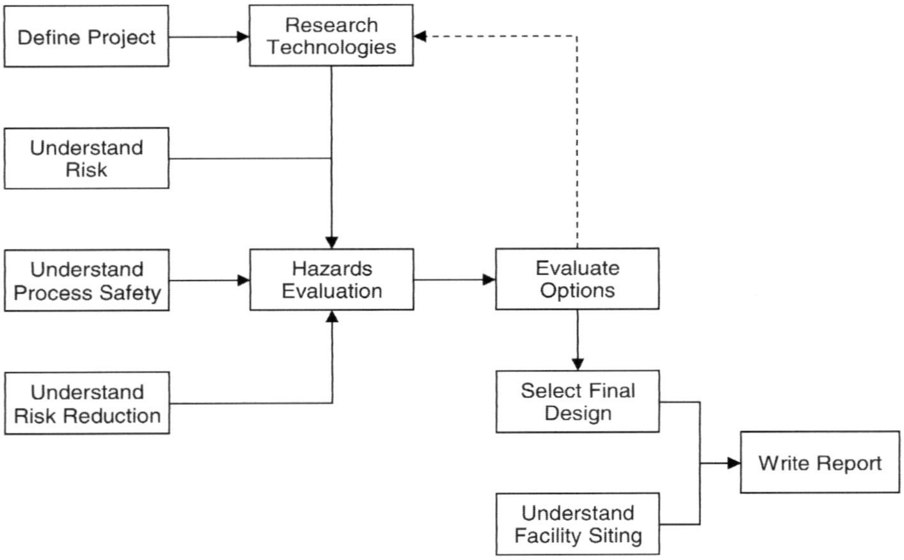

Figure 7

An Approach For Reducing Process Safety Risk

$$R = \frac{F \times C}{D}$$

R	**R**isk	
	What consequence (event) we are trying to reduce	
F	**F**requency	
	How often the event may occur	
C	**C**onsequence	
	How severe the event may be	
		Process Safety: Fires, Explosions, Toxic Releases and Runaway Reactions
D	**D**iscipline	
	How we apply our safety principles	
		Process Safety: Operational Discipline, Conduct of Operations, Human Factors

Figure 8

The Bridge Over Process Safety's Troubled Waters:

Eliminating the Gap With SAChE

(232f) Exposing the Blurry Lines Between Personal Safety and Process Safety Education: Contrasting NIOSH Prevention Through Design (PtD) with CCPS Sache

Delmar R. Morrison[1], Ryan J. Hart[1] and Pamela Heckel[2], (1)Thermal Sciences, Exponent, Inc., Lisle, IL, (2)Prevention through Design Initiative, National Institute for Occupational Safety and Health, Cincinnati, OH

Chemical engineering curricula have long been focused on instilling the fundamentals of chemical engineering science in undergraduates and advancing the state of the art through postgraduate study. Unfortunately, a comprehensive focus on safety has never been in the core of the chemical engineering curricula; however, since the T2 incident, industry stakeholders have pushed successfully for additional emphasis on process safety education in U.S. curricula. The SAChE program was created as a collaborative effort between CCPS and engineering schools to develop support materials to teach the elements of process safety to chemical engineering undergraduate and graduate students. Recently, NIOSH has led a national initiative called Prevention through Design (PtD) to address occupational safety and health needs in engineering design to "prevent or minimize the work-related hazards and risks associated with the construction, manufacture, use, maintenance, and disposal of facilities, materials, and equipment." We have found that there is a blurry line between the concepts of process safety and personal (i.e., occupational) safety in both industry and in academia. Not only is there a hole in the chemical engineering curriculum, but also the distinction between the two sets of complementary safety objectives is little appreciated. This extended abstract aims to identify the core concepts of PtD as applicable to the chemical engineering curriculum and contrast those to the typical elements of process safety. The objective of the abstract is to provide an outline for creating a PtD module for the chemical engineering curriculum that complements the SAChE modules on process safety. We anticipate significant audience feedback during the session that will assist in the development of this module and aid in its implementation.

See more of this Session: Integrating Safety Into the Chemical Engineering Curriculum - Where, What, and How (Part 1)
See more of this Group/Topical: Education

(232g) The 3rd Edition of Chemical Process Safety, Fundamentals with Applications

Daniel Crowl, Chemical Engineering, Michigan Technological University, Houghton, MI and Joseph F. Louvar, Chemical Engineering, Wayne State University, Detroit, MI

The 3rd edition of Chemical Process Safety, Fundamentals with Applications was published in the early summer of 2011. The third edition has been rewritten to include new process safety technology, new references, and updated data that have appeared since the first edition was published in 1990 and the second edition in 2002.

The third edition contains two new chapters. Chapter 8, “Chemical Reactivity,” was added due to the recommendations from the U.S. Chemical Safety Board (CSB) as a result of the T2 Laboratories accident investigation. Chapter 13, “Safety Procedures and Designs,” was added to consolidate some material that was scattered throughout the previous editions and to present a more complete and detailed discussion.

This paper will present an historical perspective on this textbook, present our original objectives for this textbook, and discuss some of the difficulties in publishing a first-time text in this area. It also includes our combined experiences of teaching process safety in both industry and academia during the past 20 years.

See more of this Session: Integrating Safety Into the Chemical Engineering Curriculum - Where, What, and How (Part 1)
See more of this Group/Topical: Education

(382a) Affecting Change: Creating a Culture of Safety within a Chemical Engineering Education Program

John F. Sandell, David W. Caspary and Anton J. Pintar, Chemical Engineering, Michigan Technological University, Houghton, MI

The goal of every safety program is to prevent accidents. However, a safety program designed for use in Chemical Engineering education should have much broader goals. It should help our students develop safety awareness and hazards recognition skills; teach safe design and operation of chemical processing activities; empower our students to establish a culture of safety within their working environment; and, improve their safety awareness in their private lives and in the communities in which they live.

The Chemical Engineering department at Michigan Technological University has 30 years of experience integrating safety education into the undergraduate curriculum. Initially, our safety program was restricted to the senior-level Unit Operations Laboratory and was an autocratic, "follow-the-rules", management-driven safety program. In the late-1980's the safety program was remodeled to reflect several successful industrial safety programs of the time and named Prevent Accidents With Safety (PAWS). In 1990, a required, senior-level course in Chemical Process Safety was added to the Chemical Engineering curriculum. In 2000, under the new ABET program evaluation criteria, the laboratory safety program was included as one of the curriculum's primary assessment tools. By the mid-2000's the PAWS safety program was implemented into the junior-level Fundamentals of Chemical Engineering Laboratory course. And in 2011, at the request of students in Unit Operations Laboratory, our regularly-scheduled safety meetings include a "safety outside the workplace" topic.

The required safety course and comprehensive PAWS laboratory safety program continue to evolve and improve. This presentation will cover important aspects of the evolution and the current scope of the Chemical Engineering safety program at Michigan Technological University.

See more of this Session: Integrating Safety Into the Chemical Engineering Curriculum - Where, What, and How (Part 2)
See more of this Group/Topical: Education

(382b) Safety Education throughout the Undergraduate and Graduate Experience At Michigan Tech

Faith A. Morrison and Adrienne Minerick, Department of Chemical Engineering, Michigan Technological University, Houghton, MI

Michigan Tech has a mature safety program in its capstone Unit Operations and Plant Operations courses (senior required laboratories; see accompanying abstract by Sandell et al.). The capstone experience builds on a comprehensive safety program that is spread throughout the chemical engineering curriculum. In this presentation, the authors describe Michigan Tech's introductory safety program, safety in the bioengineering electives, and auxiliary safety program. The introductory safety program is part of our Fundamentals of Chemical Engineering laboratory (required junior class) where we focus on establishing the seriousness of plant and laboratory safety, shared responsibility for safety, and learning to identify hazards. Specific skills addressed include appropriate use of personal protective equipment, ladder safety, use of safety eye wash and safety shower, and chemical labeling. In the bio-focused coursework laboratories, we address the special safety issues that impact biochemical engineering education. In addition to all of the chemical safety skills, students learn about routes of biocontamination and practice using engineering controls that isolate biological samples from equipment or people. Students also learn specific protocols for biosafety level 1 and 2 laboratories by working with graduate students who are trained via our graduate safety program and job safety assessment (JSA) reviews prior to setting up and conducting an experiment. This presentation will highlight the comprehensive efforts in our program to infuse students with a safety mentality.

See more of this Session: Integrating Safety Into the Chemical Engineering Curriculum Where, What, and How (Part 2)
See more of this Group/Topical: Education

A Treasure Hunt towards Process Safety In the Unit Operations Laboratory

Tuesday, October 18, 2011: 3:55 PM

Marquette I (Hilton Minneapolis)

Ronald J. Willey, Kathleen Ziemer and Tracy Carter

Northeastern University, Boston

Management buy-in helps!

- I took over the unit operations laboratory in 1986
- Ralph Buonopane was the Departmental Chair
- He supported and heavily encouraged me to integrate process safety into our unit operations laboratory

Staff must belief in safety

- Not only was I encouraged from the top
- I was pushed from the bottom
- Our technician, Al Bina, everyday, in my office asking:
    - How can we make it better?
- I make a slight suggestion, next you know…

Our laboratory utility lines are labeled

- 

Or new signage is added to throughout the Laboratory

However the biggest incremental step towards a safe culture: Our Initial Treasure hunt

- I inherited it from Ralph, who inherited it from the previous instructors
- A process that happens the very first day students are in the laboratory
- Many alumni can recall stories, and their first day experience with this process
- Continues today under Profs. Ziemer and Carter
- Prof. Carter was one of my students in the 90's

Directions to Students

- Identify equipment used in laboratory experiments (must be submitted at end of the period). Locate the following items in the Chemical Engineering Unit Operations (Baker) Laboratory. Record on this sheet the name of the item, its label [e.g. V21], and its location. Turn in one sheet per group.

Opening page

Department of Chemical Engineering
CHEM 2311 Transport 1 Laboratory

Identify equipment used in laboratory experiments (must be submitted at end of the period). Locate the following items in the Chemical Engineering Unit Operations (Baker) Laboratory. Record on this sheet the name of the item, its label [e.g. V21], and its location. Turn in one sheet per group.

1. Three types of manual valves

2. Two types of flow measurement instruments

3. Two types of level measurement instruments

4. Two types of automatic regulating valves or control valves

Portion dedicated to Safety

b. Eye wash
c. Safety manual
d. CO_2 fire extinguisher
e. Dry-type fire extinguisher
f. Lock out tag
g. Emergency first aid box
h. Fire blanket
i. Fire alarm
j. Relief valve
k. MSDS Safety sheets

Photo 1 students looking in general

- To be added before the presentation from the Sept 2011 class

Photo 2 Students looking

- to be added

Our method and the whys

- After the treasure hunt all students report back to the key instructor, generally, in the center of the laboratory.
- He or she works down the list item by item asking a student to tell where they located the particular "treasure"
- Not unusual to have even a common instrument such as pressure gauges misidentified.

An example

- Show me a check valve
- Student identifies one in the laboratory
- If incorrectly identified, we’ll explain the correct term, and if safety is involved, tie this into the explanation
- When the check valve is found, we’ll explain it’s function with a cut away

Cutaways – to many students their first experience with valves and fittings

Collectively, we review each laboratory safety item at the end with the whole group

- One key safety training item is the emergency number
- Inside Northeastern it is X 3333 (NOT 911)
- Via Cell phone it is 617-373-3333
- All students must know where fire extinguishers are located

Emergency Number Posted

Fire extinguishers

- All students must know the location of laboratory fire extinguishers.
- We also use the time to cover the different types of fire extinguishers A, B, & C
- Generally there is a student who knows the differences and we have them explain these to the students.

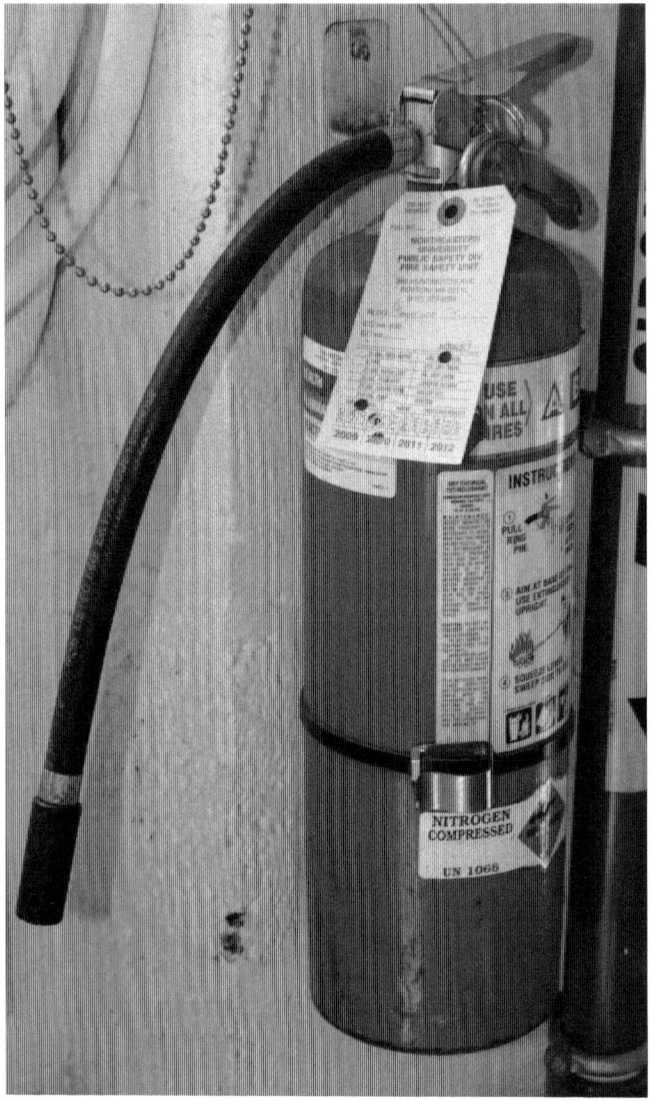

Our success? Over 25 years:

- No major incidents as defined by loss time or hospitalization
- One burn due to hot water when disconnecting a hose
- One foot ran over with a heavy piece of equipment
- One shock due to an ungrounded unit

Other ideas we stole

- Texas Tech, Robert Bethea, SACHE module on UO Lab Safety?
- Permission to start experiment – Initiated in 1989 and continues today.
- Done correctly, is extremely helpful in detecting issues beforehand and creating an awareness
- Michigan Tech University PAWS program – had about a 2 year lifetime in the early 90's one or two good suggestions evolved in that time.

Permission to Start Experiment Form

- A check list – Teaching Assistant Dependent
- AND

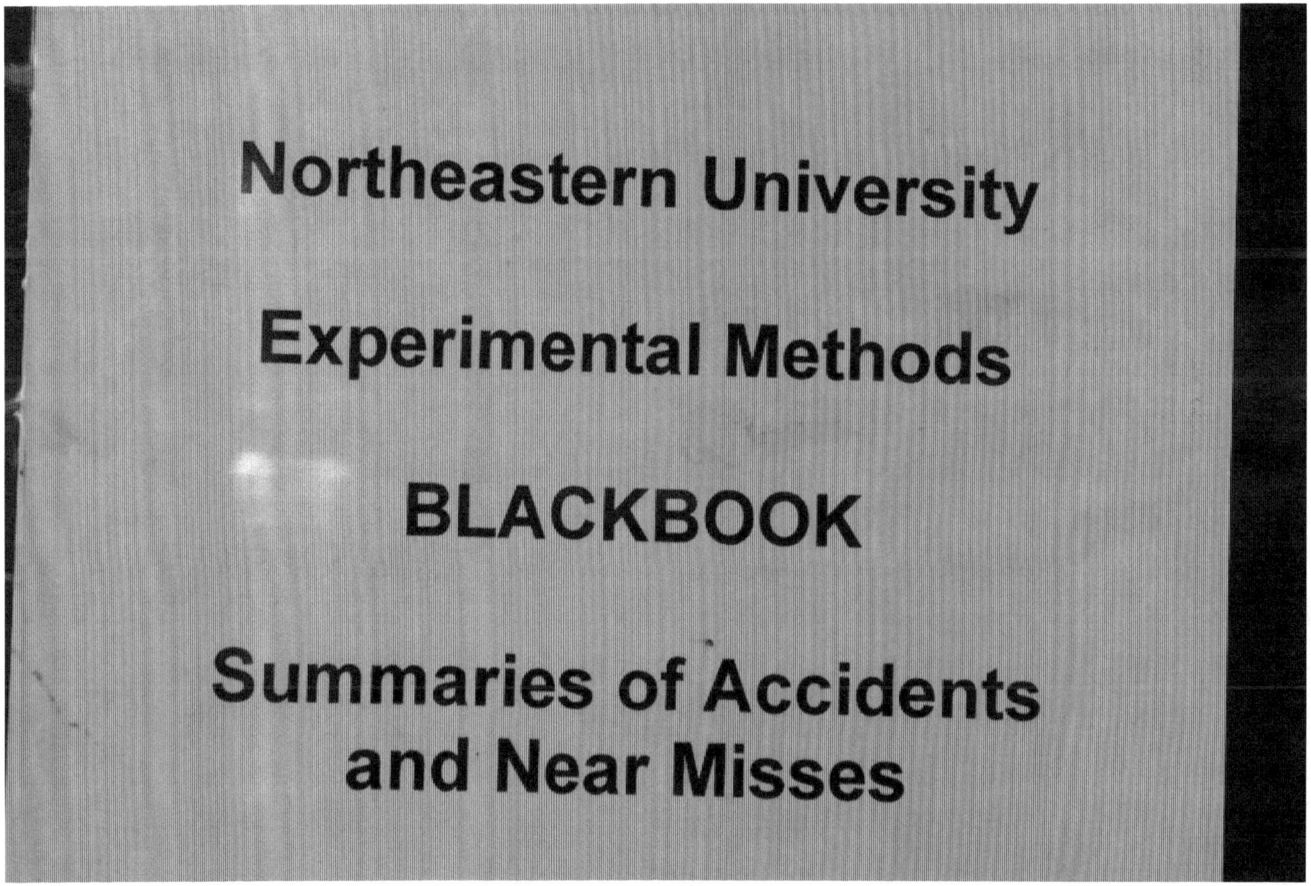

PAWS website from MTU

MichiganTech
Create the Future

Chemical Engineering

Search A to Z Map

- HOME: SAFETY PAGES
- DEPT. SAFETY MANUAL (PDF)
- MSDS DATABASES
- HAZARDOUS WASTE MANUAL (PDF)
- SAFETY FORMS
- LAB HYGIENE
- LIFTING TIPS
- PAWS SAFETY PAGE

PAWS SAFETY PAGES

- Safety Incident Reporting
- Incident Report Form (PDF File)
- PAWS Incentive Program
- Safety Committee

Prevent
Accidents
With
Safety

PAWS Safety Program in the undergraduate Chemical Engineering Labs at Michigan Tech

The purpose of the PAWS program is to develop a safety community within the students enrolled in the Chemical Engineering laboratory courses. Students assume responsibility for their own safety, for the safety of those around them, and for the safety and protection of their environment.

The goal of the PAWS program is incident and accident prevention.

Students are involved in all aspects of the safety program. All Safe Operating Procedures, the Laboratory Safety Rules, and all Process Safety Reviews are conducted and developed with active participation of our students. The Safety Manual for use in the Chemical Engineering Laboratories is reviewed and edited every year by a team of students. Students perform an annual evaluation of the PAWS program and their own safety awareness. They suggest improvements for the following year and present their findings to the department faculty in oral and written format.

MTU PAW Form Students fill oou

Prevent Accidents With Safety

Unsafe Acts:

_____Chemical Safety problem
_____Improper PPE
_____Unsafe Act in Lab or Improper Use of Equipment
_____Other: ______________________________

Equipment or Facility Problems:

_____Violation of Hazard Communication Standard
_____Leaks
_____Safety Equipment problem
_____Electrical problem
_____Faulty Equipment
_____Hot Surfaces
_____Odors
_____Missing Guards

Question to the audience

- Can you share a UO laboratory safety practice with our audience?

THE OHIO STATE UNIVERSITY
PROCESS SAFETY EDUCATION

Robert W. Johnson
Department of Chemical and Biomolecular Engineering
The Ohio State University
Columbus, Ohio

Biography

Robert W. (Bob) Johnson (B.S.Ch.E, M.S.Ch.E, Purdue University) is a Fellow of AIChE and past chair of AIChE's Safety & Health Division. In addition to being on the faculty of Ohio State part-time, he is President of Unwin Company, a process safety consultancy. He teaches the AIChE continuing education course on HAZOP Studies, and was primary author of CCPS' *Guidelines for Hazard Evaluation Procedures, Third Edition* as well as two books on chemical reactivity hazards.

Abstract

This presentation describes how process safety has been taught at The Ohio State University as a dedicated elective course, including use of the Safety and Chemical Engineering Education (SAChE) safety certificate program and other available resources.

Introduction

In addition to safety-related topics being included in the core chemical engineering curriculum at Ohio State, most notably in the Unit Ops Lab and in the senior design course, Ohio State offers a 3 credit hour Chemical Process Safety elective to Juniors, Seniors and graduate students. This elective course, originally developed and taught for two years by John E. Corn, instructor at Ohio State (now retired) and past chair of AIChE's Process Development Division, has now been taught for five years, with expanded content each year.

Text

The primary textbook for this course has been Crowl and Louvar, *Chemical Process Safety: Fundamentals with Applications* (Prentice-Hall). The process safety course used the third (2011) edition of the text for the first time last summer. The text is supplemented by a reading assignment related to Ethics and Professionalism, as well as many on-line resources.

Topics

In-class lectures are given on the following topics:

1 - Course Introduction

2 - Regulations and Ethics
3 - Safety in Process Design
4 - Fires and Explosions
5 - Fire & Explosion Prevention
6 - Dust Explosions
7 - Chemical Reactivity Hazards
8 - Introduction to Reliefs
9 - Relief Sizing
10 - Hazard and Risk Analysis
11 - Chemical Plant Security
12 - Inherently Safer Design

The course introduction is fairly extensive, covering the anatomy of process safety incidents as well as the spectrum of process hazards shown by way of videos and calculations the potential severity of consequences.

SAChE Products

Portions of several products of AIChE's Safety and Chemical Engineering Education Committee are worked into the lectures at various points, including the "Introduction to Reactive and Explosive Materials" video in the course introduction. Students are required to complete a minimum of four Safety Certificate Modules, with completion of additional modules being given extra credit. The Piper Alpha video has been shown most years, also.

Grading

Students can earn a base maximum of 100 points by the following:

10	Class attendance and participation
20	Completion of four Safety Certificate Modules
10	In-class quizzes
10	Midterm exam
21	Homework problems
14	Case study reports and presentations
15	Final exam

Extra credit points can be earned by attending more than ten lectures, completing more than four Safety Certificate Modules, and completing extensions to one of the homework problems.

Future

The Chemical Process Safety will be changing in the coming year, as Ohio State transitions from quarters to semesters. The new ABET requirements are likely to also have an impact. However, it is not known at this time what form that change will take.

IMPLEMENTING CONSERVATION OF LIFE ACROSS THE CURRICULUM

Richard A. Davis, Department of Chemical Engineering, University of Minnesota, Duluth, MN
James A. Klein DuPont, North America Operations, Wilmington, DE

Introduction

At the University of Minnesota Duluth, the Chemical Engineering Department faculty has significant industry experience that includes industry training in the practice of chemical process safety (CPS). As a faculty, we value CPS instruction in our curriculum foremost because we do not want our graduates to get hurt or hurt other people. Regardless of the nature of industries that employs our graduates, or whatever future engineering assignments or roles they may take on, there are benefits to student's exposure to CPS principles. Lessons learned from historical accidents and recent incidents remind us of the value of informing our students about their role in CPS. Starting now to develop a mindset or culture of CPS will help our graduates integrate into any organization's management of chemical process safety. The more knowledge our students have, the less likely there will be serious incidents.

We have emphasized safety in our curriculum primarily in our labs, to remain in compliance with OSHA regulations and the campus Environmental, Health, and Safety (EHS) policies, and in our capstone design course sequence. Since the mid 1990's, half of the faculty has participated in the SAChE Faculty Workshops. As a result, our faculty has incorporated a sprinkling of various safety topics in other courses at the discretion of individual faculty members. We were relatively satisfied with our level of coverage based on exit interviews with graduating seniors and survey feedback from alums. However, we began to see signs that our level of preparation for our students was no longer serving their needs. Students returning from internships and industrial cooperative education experiences were describing training and events involving CPS in the workplace. With the renewed emphasis on safety in the aftermath of relatively recent industrial accidents, such as T2, and in anticipation of ABET revisions of student outcomes, we have made a concerted effort to coordinate and assess CPS awareness and skills of our students. We also began to experience industry outreach from practicing engineers on this topic.

The main challenge we faced was how to add more CPS instruction into a full program with little room for new material. We determined that, rather than add a new required course, at the expense of technical electives credits that are popular with our students, we would instead weave safety throughout the curriculum with the aim of graduating engineers that have a culture of safety, and a base set of skills to practice engineering safely (Pintar, 1999). We have four goals:

(1) Make students aware of CPS principles, practice, and the chemical engineer's role.
(2) Infuse a culture of life-long safety in our department faculty, students and graduates.
(3) Assess student learning, basic skills, and commitment to CPS.
(4) Require minimal additional resources or student credit hours in a full curriculum.

Implementation Plan

To accomplish our goals of training our students in CPS, we embarked upon a four-step implementation plan:

1. Unify our CPS instruction across the curriculum around a common theme: Conservation of Life (COL). (Klein & Davis, 2011)
2. Survey the faculty to benchmark our current state of CPS instruction, identify holes, and add/eliminate elements to the curriculum as needed.
3. Inventory COL teaching resources for instructors and courses.

4. Develop a set of student learning outcomes and rubrics to assess student learning as part of our overall assessment activities.

As a first step towards implementation, we engaged all of our constituents in the process of planning and implementation, beginning with our department's external advisory board. Our board is primarily composed of practicing chemical engineers with a wide range of industry experience. We recruited a new member to the board who has specific CPS training and professional responsibility, to champion the plan to the board and provide additional guidance to the faculty.

We also unfolded the plan to our students through academic advisement, in our courses, and extracurricular programming, such as the student chapter of AIChE. Graduates of the program are periodically surveyed about CPS as part of our program assessment activities.

Create a Culture of Safety

The first element of process safety management (PSM) is to establish a strong culture of safety beginning with leadership and commitment from management (Klein & Vaughen, 2011). Within an academic department, our process is training students, and we might think of the department head and faculty as managers of this process. However, we also involve our students early in PSM. In our first-year introductory course, students are made aware of CPS concepts, terminology, and the role chemical engineers play. We also teach them the department approach to PSM, with our theme of COL. All students, faculty, and staff are held to high standards of operational discipline, whether teaching the COL principles of CPS in our courses, practicing Environmental, Health and Safety (EHS) activities in our teaching and research labs, or involving students in the management process.

Conservation of Life as a Unifying Theme

Klein and Davis (2011) proposed the theme of "Conservation of Life" to elevate chemical process safety instruction, principles, and practice to the same ubiquitous level as other principles of conservation of energy and mass in chemical engineering education. They developed five principles based on industry standards and practice:

1. Assess material and process hazards (develop basic data on reactivity, flammability, toxicity)
2. Evaluate hazardous events (apply methodologies to estimate potential hazardous impacts)
3. Manage process risks (evaluate risk vs. acceptable risk criteria, apply inherently safer approaches, evaluate and design multiple layers of protection)
4. Consider real-world operations (apply comprehensive PSM systems, recognize importance of human factors, learn from experience – case histories)
5. Ensure process & product sustainability (product safety stewardship, life-cycle management)

The five COL principles serve as a guide for curriculum development, coverage of topics, and implementation. We have implemented CPS curricula according to a spiral-learning model that provides students with increasing levels of depth and breadth in the five COL principles as students move from lower to upper division courses, culminating in a capstone design experience.

Table 1 identifies some of the methodologies and tools for each of the five principles of COL, as well as recommendations for placement within a chemical engineering curriculum. Instructors introduce or review CPS topics within our theme of COL to elevate the importance of safety to the level of conservation of mass and energy in engineering practice.

Table 1 Conservation of Life in the Curriculum

COL Principles	Methodologies	Tools	Where in the Curriculum
1. Assess material/process hazards	• MSDS • Chemical Hazard Characterization • Reactivity Screening	• NOAA Chemical Reactivity Worksheet - reactivity database • NOAA CAMEO Chemicals: response recommendations and reactivity prediction tool • DOW Chemical Exposure Index • ASTM CHETAH: reactive hazard screening (commercial) • EPA High Production Volume Information System (HPVIS)	• Chemistry • Intro to ChE • Materials Science • Unit Operations • ChE Labs • Reaction Engineering • Capstone Design
2. Evaluate hazardous events	• HAZOP • Root Cause Analysis	• EPA DEGADIS: dispersion Model • EPA/NOAA ALOHA: hazard release model • DOW Fire & Explosion Index	• Thermodynamics • Unit Operations • Reactor Design • Particle Tech • Capstone Design
3. Manage process risks	• LOPA • Inherently Safer Design	• EPA RMP*Comp: risk management plan	• Unit Operations • Process Control • Capstone Design
4. Consider real-world operations	• PSM • Case Studies		• Intro to ChE • Capstone Design
5. Ensure process & product sustainability	• Fate Screening	• EPA EPI Suite: screening tool for physical/chemical/environmental properties	• Capstone Design

Benchmark Current Practices and Propose Curriculum Modifications

Once we have formally identified the guiding principles of COL, we needed to establish a baseline of current CPS education in our curriculum. We surveyed the faculty to identify, with examples, where they are covering COL principles in our core courses. We recognize that many elective courses also incorporate COL principles, which is encouraged. However, we restricted our survey and implementation plan to the core set of courses required of all of our graduates. In this way, we ensure that all of our students are meeting our standard for CPS education.

Table 2 shows a matrix of a snapshot in time mapping general coverage of COL principles to our program's core courses. The faculty survey asked members to rank their level of coverage from zero = no COL to three = Advanced COL. Table 2 gives a view of where the potential holes exist in the curriculum. At first glance, it appears that we had adequate coverage with some level indicated for each principle. However, upon further inspection of examples and discussion with the faculty, we realized that the coverage was weighted towards the introductory level, lacking in depth. In our labs, for example, the focus was heavy on personal protective equipment, chemical waste disposal, and general lab safety, and light on CPS. There was also a reliance on the capstone design course to serve as a catch all for most of the topics in any depth.

Table 2 Initial benchmark survey of course content * = introductory, **= intermediate, ***= advanced.

COL Principles / Core ChE Curriculum	1. Assess material/ process hazards	2. Evaluate hazard-ous events	3. Manage process risks	4. Consider real-world operations	5. Ensure product sustainability
Introduction to Chemical Engineering	*			*	
Particle Technology	**	*		*	
Heat & Mass Transfer		*	*		
Process Control		*	*	*	
Separations			**	*	
Reactor Design	*	*		*	
Unit Operations Laboratory	*	*		*	
Capstone Design	*	***	**	*	*

As an ongoing process, the faculty meets annually for learning outcome assessment. Our discussions include COL instruction. At the start, we identified holes in the curriculum and made plans to strengthen the delivery of material and active learning exercises where needed. For example, we added reactive screening analysis to our reactor design course.

Inventory of COL Resources

A large cache of CPS instructional resources is available in the literature, on the web, from our industrial partnerships, within our community of educators, and our professional societies. As we started to build up our inventory of COL resources, we naturally gravitated towards the "low hanging fruit." The two quickest ways to get started are to require students to participate in the Safety and Chemical Engineering Education (SAChE) Student Certificate Program and invite industrial practitioners to make presentations to students. (SaChE Student Safety Certificate Program, 2010)

The SAChE Student Certificate program involves web-based delivery of content followed by an on-line assessment of student learning. The modules are self-contained and do not require any class time for students to complete, but they do them on their own, outside of class. Student members of AIChE are recognized for successfully completing a module with a certificate. Instructors may also access a report of successful student completion. This program satisfies one of our criteria by making students aware of COL principles with no additional resources other than a department's SAChE membership.

We find the industrial community ready and willing to visit our classes and work with faculty to deliver presentations and material on COL principles. An industrial safety professional works with our faculty member teaching the capstone design course to develop HAZOP active learning strategies. Although their presentations do require class time, they fill gaps in the expertise of the instructor and reinforce the value, from an industry perspective, of the principles of COL in our pursuit of a culture of safety in our graduates. In our case, industrial partners have sent individual faculty slide presentations and video from safety training that they find particularly relevant to undergraduate engineering education.

SAChE provides additional products for academic and industrial safety education and training. Many of these products include video, presentation slides, homework problems, and other tools. Table 3 lists our core classes and how we have integrated many SAChE modules, in addition to the Student Certificate program, across our curriculum. The table also shows where our faculty uses other videos and case studies in our courses. For example, instruction in technical communication is one objective of our unit operations laboratories. In a weekly lab class exercise, a student lab team is assigned to make a short, but formal, CPS presentation to the class based on a Process Safety Beacon of their choosing. We learned from a student participating in an industrial internship that she was required to make a similar presentation at a weekly safety meeting the first week on the job. She selected a Beacon on dust handling that gave the engineers cause to review how they were handling solid materials at their loading dock. Members of AIChE and most academic professionals now have access to books produced by AIChE's Center for Chemical Process Safety (CCPS) through the online resource Knovel. These books provide detailed information about COL principles, methodologies, tools, and case studies. For example, the Failure Scenario tables in Guidelines for Design Solutions for Process Equipment Failures (CCPS/AIChE, 1998) are an important resource for students applying inherently safer design in unit operations and process design.

Table 3 Materials for implementing COL across the chemical engineering curriculum.

Core Courses	SAChE Modules	Videos
Intro to ChE	• PSM & COL (2012) • Safety in the CPI (2006) • Bhopal (2010) • Mini-Case Histories (2003) • Process Safety 101 Certificate	• CSB Videos (Variety) • History, Modern Marvels, Engineering Disaster Series 10 Union Carbide Explosion
Mass & Energy Balances	• SHE for Textbooks (2003) • SAChE Seveso (2008)	
Thermodynamics	• Explosions (2009) • SHE for Textbooks (2003)	
Design of Engineering Experiments	• Risk Assessment (2008)	
Computational Methods	• Computer Applications for Process Safety	
Fluid Mechanics	• Two-Phase Flow/Pressure Relief (2011) • Consequence Modeling Source Models (2004)	
Particle Technology	• Static Electricity (2007, 2008) • Dust Explosions (2006) • Dust Explosion Control Certificate	• CSB Dust Explosion (multiple vids) • OSHA Fact Sheet & Poster
Heat & Mass Transfer	• SHE for Textbooks (2003)	

Engineering Materials	• Properties of Materials (2007) • Chemical Reactivity Hazards Certificate	
Process Control	• Design for Overpressure and Underpressure Protection (2006) • PSM Safety Valves (2003)	
Separations	• Inherently Safer Design (2006) • Packing Fires (2004)	• CSB Anatomy of a Disaster
Reaction Engineering	• Explosions (2009) • Polystyrene Reactor Runaway (2011) • Runaway Reactions Certificate	• CSB Reactive Hazards • CSB Runaway at T2
Unit Operations Laboratories	• CCPS Process Safety Beacon • Process Hazard Analysis (2009) • Project Risk Analysis (2009) • Risk Assessment Certificate	• OSHA PPE & Safety Videos
Capstone Design	• Piper Alpha (2007) • Hazop Case Study (2003) • Layer of Protection Analysis (2011) • Safety Guide for Design Projects (2011) • Risk Assessment (2008) • Venting (2007) • Green Engineering (2004) • Safety in the CPI Certificate • Inherently Safer Design Certificate	• Acceptable Risks (Pitt & Robinson, 2003)(Strum, 1986) • National Geographic Seconds from Disaster series "Texas Oil Explosion", "Explosion in the North Sea," • History, Modern Marvels, Engineering Disaster Series 6 Piper Alpha, 11 Gas Explosion, 18 BP Refinery • CSB Videos

We also visited several government and professional society web sites for additional materials and information on CPS that supports the five principles of COL. Much of the material is available at no cost and ranges from simple posters for safe dust handling from OSHA and videos of incident investigation from the US Chemical Safety Board (CSB), to software tools for dispersion modeling from NOAA and EPA, as listed in Table 1. In another example, NOAA has a web-guided tour entitled "Responding to a Chemical Spill." The American Chemical Council maintains a database helpful for COL principle 5 dealing with product safety and sustainability.

Table 4 Map of COL principles to SAChE Recommendations for ABET Safety Content in Chemical Engineering

COL Principles \ SAChE ABET Safety Content	1. Understand importance of CPS and required resources/commitment	2. Characterize hazards associated with chemicals/agents	3. Understand/apply concepts of inherently safer design.	4. Understand how to control/mitigate hazards to prevent accidents	5. Familiar with major regulations	6. Understand consequences of incidents due to acute/chronic releas-	7. Proficient with hazard identification procedure	8. Introduction to the process of hazard evaluation/risk as-
1. Assess material / process hazards		■					■	
2. Evaluate hazardous events							■	■
3. Manage process risks			■	■	■			■
4. Consider real-world operations	■				■	■		
5. Ensure product sustainability		■			■	■		

Assessment of Student Learning

We have an assessment process already in place with tools and measures identified for meeting ABET prescribed learning outcomes. In keeping with our guiding principle requiring no new resources, we found that we are able to use our existing process to assess student-learning outcomes for COL. Before graduation, students are expected to demonstrate their knowledge and application of COL principles in their reports and presentations for our capstone design course by addressing subjects. Table 4 maps our principles of COL to the SAChE recommendations for ABET Safety content in Chemical Engineering (SAChE). However, COL goes beyond SAChE recommendations with the addition of product safety and sustainability. We adopted five learning outcomes and developed rubrics for assessing student achievement following best practices in our rubric design (Acquisition, 2010). As shown in Table 5, each rubric has four levels, with the highest level of achievement at the left. The criteria are cumulative as we move from lower to higher levels of achievement. We use our existing measurements of student achievement to assess COL education of our graduates. We provide student with these rubrics in our introductory course, and review them periodically throughout the curriculum to develop the sense of personal responsibility in the student that contributes to our culture of CPS.

Table 5 Rubrics for assessment of COL student learning outcomes. Levels are cumulative.

Outcome 1: *Graduate is able to assess material and process hazards.*

3. Exceeds Standard	**2. Meets Standard**	**1. Making Progress**	**0. Not there yet**
• Reactivity screening	• Calculate compositions relative to Lower and upper flammability and exposure limits	• References MSDS	• Minimal or no data developed

Outcome 2: *Graduate is able to evaluate hazardous events.*

3. Exceeds Standard	2. Meets Standard	1. Making Progress	0. Not there yet
• Use Fault/Event Tree Analysis, FMEA, Human Error Analysis, etc • Dispersion modeling	• Use HAZOP to identify hazards for a unit operation	• Uses a simpler "What if" technique to identify hazards	• Minimal or No identification of hazards

Outcome 3: *Graduate is able to manage process risks.*

3. Exceeds Standard	2. Meets Standard	1. Making Progress	0. Not there yet
• Performs risk analysis • Management of change applied to any modification in design • Evaluate alternative designs for inherent safety • Design involves intensification, substitution, attenuation, limitation	• Qualitative risk assessment (LOPA) • Written operating procedures • Pre-start-up safety review • Designs a unit operation/process considering multiple layers of protection	• Considers worst case scenario • Provides process safety information • Process Hazard Analysis • Identifies at least one layer of protection • Identifies opportunities for inherent safety	• No evaluation of risk • No PSM included in reporting • Does not consider inherent safety or protection in design of unit operation/process • Does not consider inherent safety

Outcome 4: *Graduate understands real-world operations.*

3. Exceeds Standard	2. Meets Standard	1. Making Progress	0. Not there yet
• Apply comprehensive PSM systems, recognize importance of human factors, learn from experience – case histories	• Process design demonstrates compliance with relevant OSHA regulations • General understanding of causes/consequences of major incidents	• Cites OSHA regulations • Aware of one incident	• No reference to regulations • Cannot cite an industrial incident

Outcome 5: *Graduate is aware of principles of product sustainability.*

3. Exceeds Standard	2. Meets Standard	1. Making Progress	0. Not there yet
• Performs a Lifecycle inventory analysis • System optimization • Considers Ultimate use and disposition of products and manufacturing materials	• Holistic design for energy and mass integration • Preliminary Environmental Impact Study • Consider product safety issues for downstream users	• Mass & energy balances closed	• Mass and energy balances not closed

Results

We are entering the third academic year of our COL implementation effort. Table 6 shows our current state of COL in the curriculum. We find it helpful to keep COL in front of the faculty by reviewing the plan and activities with the external advisory board and assessment team. COL is a common agenda item in faculty meetings.

Table 6 Current map of course content to COL * = introductory, **= intermediate, ***= advanced.

COL Principles / Core ChE Curriculum	1. Assess material/ process hazards	2. Evaluate hazardous events	3. Manage process risks	4. Consider real-world operations	5. Ensure product sustainability
Introduction to Chemical Engineering	*	*		*	*
Mass & Energy Balances	*				**
Thermodynamics	**			*	
Design of Engineering Experiments		*	*	*	*
Computational Methods	*	*	*	*	
Fluid Mechanics	*		**	**	
Particle Technology	**	*		***	
Heat & Mass Transfer	*	*	**	**	*
Engineering Materials	*		*	*	*
Process Control		**	**	***	
Separations		*	**	**	*
Reactor Design	**	**	**	***	
Unit Operations Laboratory	**	**		**	*
Capstone Design	**	***	***	***	**

Exit interviews between the department head and graduating seniors provide direct evidence that we are approaching our goal of developing a culture of process safety in the program and graduates. When asked to identify current issues in chemical engineering, 100% of the students interviewed mentioned their role and stewardship in process and product safety.

Alumni surveys indicate that our graduates value their education in COL principles, ranking these topics high in level of importance, and providing comments for inclusion and improvement.

The faculty has not wavered in its commitment to our theme of COL across the curriculum. One issue we have identified is the need for consistency between the same courses delivered by different instructors. We are resolving this issue through our planning efforts, inventory of materials, and designation of courses where materials are used in the program. This is also helping to avoid unnecessary duplication of effort and coverage. A second issue involves timing of COL coverage in a course. The faculty has learned to introduce COL principles early in a course to allow for their integration into the various topics, rather than an add-on at the end given time constraints. By including COL principles at the same level as other conservation principles, students are learning to consider safety in aspects of problems solving and design. We plan to continue to explore and develop new COL methodologies, tools, case studies, etc. to continue to enhance CPS instruction in our courses. Our external advisory board strongly endorses our COL theme and coverage of topics. The members of the board are an integral part of our refining process as we move forward.

Recommendations and Conclusions

Our experience with COL as a unifying theme has been positive and productive. The guiding principles of COL provide faculty with a framework for program development. Education resources are distributed across the curriculum in a fashion that provides complete coverage of material while giving students increasing depth of experience through multiple opportunities to practice using established methodologies. Programs may adopt this approach without significant demands of time and resources on the department and faculty. By holding students and faculty accountable through an assessment process, a program may quickly establish a culture of chemical process safety and meet required learning outcomes. Ultimately, CPS/COL in the curriculum has value for all constituents of a chemical engineering program, including the faculty, staff, students, industry, and public they serve.

References

1. Acquisition, C. C. (2010, July 15). Evaluation: Process: Creating Rubrics. Retrieved June 23, 2011, from University of Minnesota: http://www.carla.umn.edu/assessment/vac/evaluation/p_7.html
2. CCPS/AIChE. (1998). Guidelines for Design Solutions for Process Equipment Failures. New York: Wiley.
3. Crowl, D. A., & Louvar, J. F. (1990). Chemical Process Safety: Fundamentals with Applications. Upper Saddle River: Prentice Hall.
4. Jimenez-Gonsalez, C., & Constable, D. J. (2011). Green Chemistry and Engineering: A Practical Design Approach. Hoboken: Wiley.
5. Klein, J. A., & Vaughen, B. K. (2011, June). Implement an Operational Discipline Program to Improve Plant Process Safety. Chemical Engineering Progress , 48-52.
6. Klein, J., & Davis, R. (2011). Journal of Chemical Engineering Education .
7. Pintar, A. J. (1999). Teaching Chemical Process Safety: A Separate Course versus Integration inot Exisitng Courses. ASEE Conference Proceedings. Session 3213. Charlotte: ASEE.
8. Pitt, M. J., & Robinson, J. E. (2003). Using a Commercial Movie for an Educational Experience. Chemical Engineering Education , 37 (2), 154-155.
9. SAChE. (n.d.). Retrieved Auguest 19, 2011, from SAChE: http://www.sache.org/
10. SaChE Student Safety Certificate Program. (2010). Retrieved July 21, 2011, from SAChE: http://sache.org/student_certificate_program.asp
11. Goodwin, R. (Producer), Strum, N. (Writer), & Wallace, R. (Director). (1986). Acceptable Risks [Motion Picture]. USA.
12. Turton, R., Bailie, R. C., Whiting, W. B., & Shaeiwitz, J. A. (2009). Analysis, Synthesis, and Design of Chemical Processes (3rd Edition ed.). Upper Saddle River: Prentice Hall.

TEACHING SAFETY – A BRITISH AND EUROPEAN EXPERIENCE

Martin J. Pitt, University of Sheffield, Sheffield, UK

Biography

Martin Pitt worked in industry as a project engineer and chemical plant manager before becoming an academic at Leeds then Sheffield universities, UK. He formerly ran safety courses for J. T. Baker Chemicals, and continues to lecture on courses for industry. He is a long-standing member of the IChemE Safety & Loss Prevention Subject Group committee, and past Chairman of the Education Subject Group, and has run workshops on safety teaching. He is currently Chairman of the European Federation of Chemical Engineering Working Party on Education.

Introduction

Safety has long been an absolute requirement for UK chemical engineering courses accredited by the Institution of Chemical Engineers (IChemE), but has not been included in many European degrees. The IChemE Safety and Loss Prevention Subject Group arranges periodic workshops for lecturers, and the most recent one (2010) was a European one in association with the European Federation of Chemical Engineering, following increasing interest[1]. This paper is based on the author's experience as a lecturer, workshop organizer and accreditor of degree schemes.

What Should Be Taught?

All UK departments teach safety, but how useful is what they teach? The IChemE identified 28 possible topics and surveyed more than 376 graduates from 29 UK universities and 15 others. The most relevant subjects (more than 90%) were reported to be[2]:

Hazard Identification (99%); Hazop (98%); Case Studies (98%); Health and Environment (96%); Terminology (96%); Inherent Safety (95%); Safety Culture (95%); Permit to Work (95%); Risk Perception (95%); COSHH [UK legislation: Control Of Substances Hazardous to Health] (93%); Benefit Analysis (92%); Fires and Explosions (91%); Protective Systems (91%); Human Factors & Ergonomics (91%).

However, we should not be limited by popularity. It is probably true that most chemical engineering graduates will not have responsibility for chemical reactors with the potential for runaway (as most chemical engineering is not about chemistry), but if they are not made aware of the hazard potential in university, where else will they learn it? This is particularly relevant for smaller companies where the graduate may not have much technical support.

Who Should Teach It?

A practical problem is that few UK departments are active in safety research and therefore most do not have faculty with the specialist knowledge. Moreover, safety is actually one of the hardest subjects to teach since it requires an overview and appreciation of the interaction of various processes, human and technical.

In Europe, the 2001 explosion at the AZF factory in Toulouse has prompted efforts and staff commitments for safety teaching in France[3] and there has been general increase in interest in other countries following other incidents.

How to Teach

It is necessary to be realistic in deciding what can be achieved with the time and resources available. Achievements may be specified as *awareness*, *understanding* and *competence*. Awareness of both hazards and possible methods is probably the most valuable, since the graduate may at least recognise the situation. Understanding is hard for people (staff and students) with no experience of industry, and who are used to dealing with narrow questions. A general competence in chemical engineering fundamentals and commitment to being careful is probably more likely to be effective than specific skills in hazard calculations. However, quantification can be helpful in appreciating the magnitude of effects, providing care is taken to stress the uncertainty of most numbers produced.

The methods which have been used include:

Safety integration throughout the degree by inclusion in most other modules
Badged "Safety" modules (including health and possibly environment)
Teaching along with a few modules, such as laboratory, management, design
Visiting industrial speakers
A "Roadshow" of visiting academics and industrialists
Videos, computer and Internet resources

The problem is therefore not just what to teach, but who will and how. The British experience (good and bad) and some European developments may be of interest to others.

References

1. Kakosimos, K. E., and Mihailidi, E., (2010) "EFCE News and Events: European Workshop on Teaching Safety in Chemical Engineering," *Chem. Eng. Research & Design* **88** (41) iii – v.

2. Turney, R., (2006) "Which topics in process safety are valued following graduation. A survey of recent graduates;" Paper I2.1 *CHISA 17th International Congress of Chemical and Process Engineering, Prague 27-31 August 2006, EFCE Event 663*

3. Perrin, L., and Laurent, A., (2008) "Current situation and future implementation of safety curricula for chemical engineering education in France" *Educ. Chem. Eng.* **3** (2) e84-e91

AUTHOR INDEX

American Institute of Chemical Engineers
3 Park Avenue
New York, New York 10016-5991

CURRAN ASSOCIATES INC
proceedings.com